Yvon Régis Lazare Mameleyen Ada

Trade in motor pumps for small-scale irrigation in Burkina Faso

Yvon Régis Lazare Mameleyen Ada

Trade in motor pumps for small-scale irrigation in Burkina Faso

Real operation, technical characteristics of the equipment, reality of the proposed after-sales service

ScienciaScripts

Imprint

Any brand names and product names mentioned in this book are subject to trademark, brand or patent protection and are trademarks or registered trademarks of their respective holders. The use of brand names, product names, common names, trade names, product descriptions etc. even without a particular marking in this work is in no way to be construed to mean that such names may be regarded as unrestricted in respect of trademark and brand protection legislation and could thus be used by anyone.

Cover image: www.ingimage.com

This book is a translation from the original published under ISBN 978-620-3-43806-2.

Publisher:
Sciencia Scripts
is a trademark of
Dodo Books Indian Ocean Ltd. and OmniScriptum S.R.L Publishing group
Str. Armeneasca 28/1, office 1, Chisinau MD-2012, Republic of Moldova, Europe
Printed at: see last page
ISBN: 978-620-5-36004-0

Copyright © Yvon Régis Lazare Mameleyen Ada
Copyright © 2022 Dodo Books Indian Ocean Ltd. and OmniScriptum S.R.L Publishing group

TABLE OF CONTENTS

<u>DEDICATION</u>

A

GOD, the almighty father,
thanks to whom everything is possible;

My late mother, Marie Pascaline MACKOSSI ;

My father Michel MAMELEYEN ;

My brothers and sisters;

My daughter Chancy Espérance ;

I dedicate this thesis to you!

<u>ACKNOWLEDGEMENTS</u>

This report is not only the result of three months of work, but of all the years spent on the school bench. These few lines will never allow me to thank all those who, directly or indirectly, have contributed to it, whatever the nature of their contribution. May they find here all the expression of my deep gratitude.

I will mention here a few names and all those who do not have their own in this list will undoubtedly recognize them and will certainly understand that this is only due to the limited volume that I would like to give to this report.

I would therefore like to thank :

- My supervisors from EIER and ARID: Moussa Laurent COMPAORE, Jean-Paul LUC and Hervé OUEDRAOGO, for all the preliminary work they did with me, especially for the advice and suggestions they gave me before leaving for the field and throughout this work;

- We would also like to thank the PPIV Coordinator, Mr. Alphonse OUEDRAOGO, for providing us with a means of travel during our field missions;

- Mr. Seydou SANA, Mr. Amidou SAVADOGO, and Mr. Oumar TRAORE for their total availability, for all the advices they gave me, for all the means they put at my disposal during the documentary research, our field trip and during the writing of the thesis;

- All the teachers, leaders and staff of EIER for the three years spent together in this institution;

- All my schoolmates, those of the EIER with whom I spent many pleasant and unforgettable moments as well on the work level as on the social level;

- I would also like to thank all my compatriots and friends in Burkina Faso who took the place of my family during my studies in Ouagadougou, without forgetting my dearest Prudence for her moral support, the OUEDRAGO family and Mr. Mathurin B. ;

- All my family who always knew how to give me their trust and patience and thus allowed me to have a reason to continue working whatever the adversity;

- I particularly thank my father Michel MAMELEYEN without whose help I cannot imagine what I would have become;

- Finally all those who helped me during my studies;

-	I pray to Heaven that He may grant you His blessing!

AUTHOR : Yvon Régis Lazare MAMELEYEN ADA N'GOZON

Framers:
Hosting structures :
Jean Paul LUC
M.L. COMPAORE
ARID- APPIA
Hervé OUEDRAOGO

PPIV

THEME

Formal and informal trade in motor pumps for small-scale irrigation: actual operation, technical characteristics of the equipment, reality of the after-sales service offered.

SUMMARY:

The development of trade in motor-driven irrigation pumps is considered one of the main causes of the development of irrigated agriculture, which has become essential in Burkina Faso to ensure the stability of agricultural production.

The increasing demand for vegetable products by urban and rural populations is one of the main factors contributing to the success of motorized irrigation.

This study, which focuses on the market for motor pumps, their operation and technical characteristics, provides a set of answers and lessons learned from surveys and observations in the field.

The diagnosis of the functioning of the pumps allowed us to see the state of use of these materials on the study sites located in the departments of Loumbila, Mogtedo and around Lake Bam in Kongoussi. Then it was a question of listing and identifying the different makes and models of motor pumps, their characteristics as well as their purchase costs, studying their performance and proposing remedies to the problems posed: the motor pumps distributed formally and informally and installed on the irrigated perimeters are poorly dimensioned and there are also difficulties in acquiring spare parts and maintenance

The proposed sizing method makes it possible to choose, in a technical and economic way, an adequate material according to the surface and the needs of the crops.

Key words: motor pumps, Burkina Faso, irrigation, operation, technical characteristics, diagnosis.

<u>FOREWORD</u>

In order to obtain the diploma of rural equipment engineer, which sanctions the end of the studies in initial training at the Inter-State School of Rural Equipment Engineers (E.I.E.R), the third quarter of the third year is essentially devoted to the work of dissertation on a topic chosen by the student engineer according to his centers of interest and which he

It is within this framework that we dealt with the theme: "*Formal and informal trade in motor pumps for small-scale irrigation: actual operation, technical characteristics of this equipment, reality of after-sales service*" proposed by the Regional Association for Irrigation and Drainage (ARID).

The theme on which our study is based is surveys and observations of users and traders of pump sets.

This study allowed us to put into practice the theoretical disciplines and tools that were taught. It also gave us the opportunity to familiarize ourselves with data collection techniques that can be used to analyze and understand a problem.

<u>INTRODUCTION</u>

CHAPTER 1

I.   BACKGROUND AND RATIONALE

In Burkina Faso, as in many countries in the Sahel where water is often scarce, the development of irrigation requires water control in order to improve productivity and guarantee food security. To make up for agricultural production deficits (due to insufficient rainfall), and to meet the needs of plants, farmers are resorting to increasingly motorized and efficient means of water extraction. The need to use motorized pumps for the expansion of irrigated perimeters is obvious.

Motor pumps have become increasingly successful because several factors have contributed to their diffusion, notably the technological progress made over the last few decades, the sharp drop in purchase prices, which favors access to the market for small irrigators, and the specific context of Burkina Faso with the construction of numerous dams, some of which are very large and have increased water availability.

Permanent sales channels have been set up and competition is felt between the formal and informal sectors. It has been observed that most of the equipment sold in the formal and informal trade does not meet the needs of the market and that there are difficulties in providing after-sales service. The proper functioning of this equipment depends on the quality of the parts that make it up (technological factor), their characteristics and their performance in order to meet the needs and operating conditions.

This study deals with the specific case of small-scale irrigation and consists of presenting a set of answers drawn from surveys and field observation. At the end of this study, the aim is to understand the main technical and economic aspects of pumping equipment.

CHAPTER 2

II. OBJECTIVES OF THE STUDY

11.1. <u>Main objective</u>

The main objective of this study is to characterize the trade circuits of motor pumps and to make an inventory of the functioning of the equipment in the field.

11.2. <u>Specific objectives</u>

Around this global objective it will be question of :

- Identify the brands of motor pumps, note their characteristics and diagnose their actual operation in order to identify existing problems;

- highlight the technical characteristics of the equipment, their operating performance for the development of irrigation;

- to collect the concerns of farmers in order to make proposals for improving the conditions for the development of small-scale irrigation;

- To identify the appropriate solutions for an adequate use of these equipments and to determine the reality of the after-sales service to define the actions for a better maintenance and a correct upkeep of the material.

11.3. <u>Results</u>

The main results achieved as a result of this study are:

- a diagnostic analysis of the actual functioning of the equipment and facilities;

- an interaction between the operation of these pumping equipments and the operations;

- Implementation of concrete and reliable proposals concerning the choice and technical management of this equipment, its maintenance and after-sales service.

CHAPTER 3

III.   STRUCTURING OF THE REPORT

The report is divided into three (3) distinct parts:

Part A deals with pumping equipment: the motor pump. It gives a brief presentation of the study areas and the location of the perimeters that served as study sites. It also presents a technical and technological description of the motor pump, the different types, the elements of formal and informal trade and the purchase costs;

In part B we have made a diagnostic analysis of the functioning of the equipment on the basis of the surveys carried out in the field and identify the existing problems. We have also presented the different results of the operating tests carried out in the field.

The last part deals with the interaction between pumping equipment and farm characteristics. The cultivation practices, the description of the fields, the management and the irrigation techniques have been exposed. It also gives the results of field measurements.

We then made proposals for solutions based on the diagnosis of the operation:

- A sizing proposal that allows you to choose a motor pump in accordance with your needs;
- the need for good maintenance and after-sales service, which are essential elements to ensure a better exploitation of the equipment while guaranteeing a much longer life span;

CHAPTER 4

IV. PRESENTATION OF THE STUDY AREAS

The areas covered by the farmer surveys are located around the city of Ouagadougou. They are basically agricultural activity zones administratively attached to the provinces of Ganzourgou, Oubritenga and Bam.

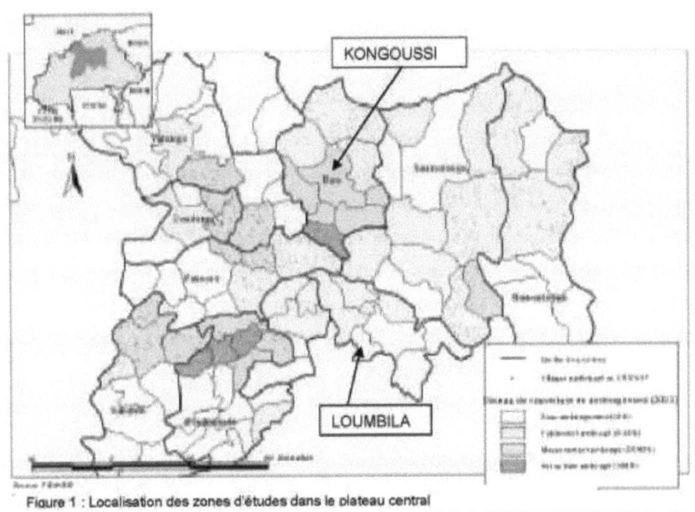

Figure 1 : Localisation des zones d'études dans le plateau central

IV.   1. <u>Commune of Kongoussi</u>

IV.   1.1 <u>Geographical location</u>

Located 110 km north of Ouagadougou, between the parallels 13° 19 50 of North latitude and the meridians 01°03 ' of West longitude, the city of Kongoussi is the chief town of the Province of Bam.

IV.1.2. <u>Climate and vegetation</u>

The commune of Kongoussi is subject to the tropical Sudano-Sahelian climate characteristic of the entire province. This climate results in the alternation of two seasons: An eight-month dry season between October and May, and a four-month rainy season from June to September. The dry season is marked by the harmattan which is a hot and dry wind often laden with dust. The rainy season is marked by the pseudo-monsoon regime, which is a wind from the West or South.

West loaded with moisture and giving rise to precipitation. In general, the rainfall is characterized by its weakness and its irregularity (500 to 600 mm per year).

The vegetation of the Commune of Kongoussi is essentially characterized by a shrubby savanna evolving towards the steppe. There is a progressive degradation of the vegetation cover linked to climatic hazards and anthropic action. This situation also affects the hydrography.

IV.1.3. *Hydrography*

The hydrography of the Commune of Kongoussi is essentially characterized by the presence of Lake Bam. With a variable length of 15 to 25 km and a width of between 200 and 600 m in the dry season, Lake Bam has its source 45 km north of Kongoussi in the department of Bourzanga. The lake's watershed covers approximately 2,600 km^2 . At the time of major floods, the length of the lake can reach 40 km with a width of 1 km, thus feeding the Dem and Sian lakes in the Sanmatenga province. The lake constitutes a permanent natural water reservoir that favors the practice of winter and off-season crops. It is also used for watering animals. The lake plays an important role in the local and national economy.

IV.1.4. *Population*

The resident population in the commune of Kongoussi was estimated at 10,559 according to the results of the 1985 general population and housing census. In the 1996 population census, the city of Kongoussi had 17,893 inhabitants, or 18.5% of the departmental population. Between 1985 and 1996, the growth rate of the communal population was 4.4%. Based on the growth rate, the population of the commune in 2004 is estimated at 26,362.

IV.1.5. *Agriculture*

In the province of Bam two types of agriculture are present:

1°) rain-fed agriculture which absorbs a large part of the active population. It is an economic activity of great importance. The main speculations of rain-fed agriculture are cereal products;

2°) Irrigated agriculture: this is the second economic activity (off-season activities) that makes up for the cereal deficits.

Market gardening is a highly developed activity in the commune of Kongoussi. Market gardening absorbs a large part of the communal population and provides substantial income. Green beans, onions, tomatoes, eggplants and cabbage are the main crops. The market garden products are strongly

marketed and even exported to neighboring cities in the country. They provide substantial income to producers.

IV.   1.6. *Soils*

The commune of Kongoussi has a great diversity of soils:

- Lithosols on ferruginous or bauxitic cuirasses are the most extensive and are located on the Birrimian hills to the west and east of Kongoussi. These are poorly developed, shallow soils (<45 cm) with little or no agricultural value.

- Tropical ferruginous soils are located on the upper slopes of the hills in the northern part of the commune. They are shallow (<45 cm) and come from the alteration of the cuirasses.

- The sandy-loam to clay soils are located along the lake in the valleys or valley bottoms. These are deep soils (> 40cm) with an interesting agro-sylvo-pastoral value.

In addition to these three main soil units, there are scattered lithosols of the cuirassed plateaus.

IV.   1.7. *Socio-economic activities*

Economic activities in the commune of Kongoussi are based on three sectors of activity: primary sector activities, secondary sector activities and tertiary sector activities.

The primary sector occupies a predominant place in the national economy. Indeed, it absorbs more than three quarters of the active population and contributes to more than 80% of the gross domestic product. In rural areas, more than half of the rural population earns its income from this sector.

IV.   2. **Mogtedo**

Geographic location

Mogtedo is a department of the Ganzourgou province located 85 km from Ouagadougou on the Ouagadougou-Fada axis, road no. 4, 25 km from Zorgho, capital of the Ganzourgou province.

The geographical coordinates of the site are :

- longitude : 00° 50' West

-latitude: 12° 18' North

- altitude : 272 m

IV.2.1. *Climate and vegetation*

Mogtedo is in a North Sudanese climatic zone with anthropogenic savanna vegetation made up of Sudanese and sub-Saharan species. The town of Mogtedo is located in a region with an average annual rainfall of about 720 mm. The average daily temperature varies between 25° and 33°C.

IV.   2.2. *Population*

The population of Mogtedo was estimated in 1985 (source: INSD) at 7056 inhabitants, the majority of whom were Mossis and a minority of Peulhs.

IV.   2.3. *Socio-economic activities*

Today, Mogtedo is the largest commercial center in the province, a significant asset of the irrigated perimeter whose crops (rice and market garden produce) are largely exported to neighboring countries (Togo, Niger and Ghana).

IV.3 Loumbila

Geographic location

Loumbila is a department in the province of Oubritenga, located about 20 kilometers northeast of Ouagadougou, the capital of Burkina Faso.

The geographical coordinates of the site are :

- longitude : 01° 24' West

-latitude :   12° 29' North

- altitude : 278 m

IV.3.1. *Climate and vegetation*

Loumbila has the same climatic and vegetative characteristics as the capital Ouagadougou. The climate of the area is marked by two clearly defined seasons. The first, dry, runs from October to April; the second, rainy and hot, from May to October. The average annual temperature is calculated at 28.2°C.

The vegetation of the locality is shrub and tree savannah consisting of shrubs, short grasses and patches of bare or sparsely covered soil.

IV.3.2. *Hydrography*

The Loumbila dam watershed is located in central Burkina Faso, between 1°18'36" and 1°54'43" west longitude, and 12°25'30" and 12°49'4" north latitude.

It is among the most populous rural regions of Burkina Faso. With an area of approximately 2,120 km^2 , its watershed extends over the provinces of Kadiogo, Oubritenga, Boulkiemdé and Kourwéogo. It is located between isohyets 700 mm and 1,000 mm.

IV.3.3. *Agriculture*

Around this dam, an intense market gardening activity has developed. In fact, dozens of associations and groups of women and young people exploit several hectares of irrigated areas.

CHAPTER 5

V.   METHODOLOGY

In order to achieve the objectives and respond to the various needs specified in the terms of reference, the actions carried out required the establishment of a simple but rigorous approach methodology, as the time and means available were limited.

This study was carried out in the following stages:

V. 1. Documentary research

The understanding of the subject and the analysis of the terms of reference are key points in the conduct of this study. A very thorough documentary research was carried out in order to deepen the knowledge of the subject. It allowed us to collect the available bibliographic references and to make an inventory of the existing data and information.

This information was sought:

- in the documents available at the library;

- on the Internet;

- with the resource persons of the structures.

Other documents related to previous or current studies were also reviewed.

V.2. Field visits

These visits allowed us to identify and locate the different sites, the farms present there and the different aspects of the work, to make contact with the farmers and to know the spatial organization of the perimeters.

They also allowed us to evaluate the extent of the work required, and to estimate the time needed to achieve the results and to set a study schedule.

V. 3. Surveys and fieldwork:

The field surveys with farmers were conducted in three (3) stages:

V.   *3.1. Selection of survey areas*

Three main criteria were used to select the areas:

- The first is the nature of the water point on which the motor pump is installed (well or surface water).

The wells have very limited flows and the crops are intensive, whereas the ponds or rivers have small amounts of water and the fields are more extensive;

- The second criterion is the location of farms in relation to marketing opportunities. We sought to identify differences in farmers' practices according to their distance from urban or local markets;

- The third criterion is the type of organization of farmers on the farms.
We also measured gaps in mechanical maintenance and spare parts supply.

Table 1: Survey areas selected

Zones	Locations	Surface water	Well
Rural	Talembika	X	
Close to a city	Loumbila	X	X
Strong commercialization	Mogtedo and Kongoussi	X	

V.3.2. *Surveys of farmers*

From a methodological point of view, we proceeded from the first week to a kind of contact with the target groups, leaving each one free to set the date of the meeting.

In all the locations surveyed, about 15 farms were selected according to the criteria defined above.

Each farmer was visited by the investigator. The individual interviews were conducted using a questionnaire on the motor pumps, the sites, the crops grown and their surface areas, the method of irrigation, the economic aspects (inputs, revenue), the limiting factors of the market gardening activity and the after-sales service offered by the suppliers. These field surveys lasted two (2) weeks.

The responses to these questionnaires were reliable in that we were able to discuss the content with the farmers before recording them.

V.3.3. *Survey of motor pump traders and repairers*

This third stage involved a limited sample (depending on the model of motor pumps) and concerned the main suppliers and repairers of these materials, and small traders in the informal sector.

Various interviews with suppliers allowed us to discuss the real market, the brands and the purchase costs of motor pumps, the supply of spare parts and the services offered to customers. This part of the survey lasted one week.

V. 4. Field measurements

They mainly concerned flow rates, pumping geometrical heights, fuel consumption and finally the determination of the HMT and efficiency. This allowed us to record the specifications of the pumps and to verify their performance in order to better diagnose their operation. The machines tested were loaned by farmers and vendors in Ouagadougou. The materials and means used were

- the decameter for surface measurements;

- the water level for the discharge heads (Hg_{ref});

- a stopwatch and a 200-liter canister for flow measurements.

V.5. Data analysis and processing

The purpose of this phase was to analyze and interpret the data from the literature review and the field work. The various data collected in the field were used first to establish a diagnosis of the operation of the pumping equipment. The analysis of the survey forms was done in two steps with Excel software and were interpreted on a graphical basis.

In the first step, it was necessary to create a data entry mask that took into account the information sought and the data collected from each card during the surveys. Then we proceeded to cross-reference the data entered to find the relevant information. The most reliable results were retained, and outliers were eliminated.

V.6. Work schedule

The figure below is a simplified view of the different phases that will punctuate the work to be done in this study.

Figure 2: Work chronogram

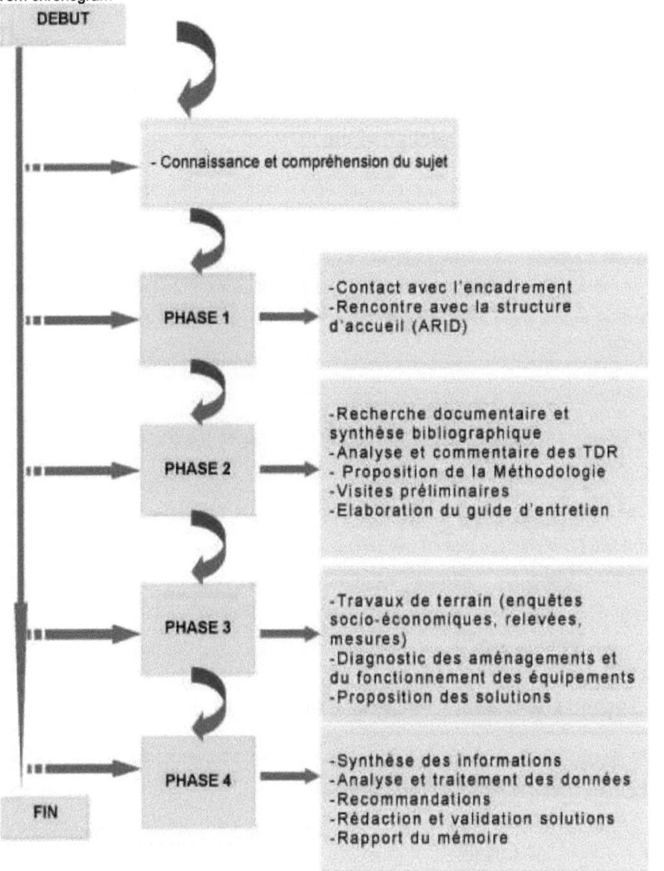

DEBUT
- Connaissance et compréhension du sujet
PHASE 1
-Contact avec l'encadrement
-Rencontre avec la structure d'accueil (ARID)
PHASE 2
-Recherche documentaire et synthèse bibliographique
-Analyse et commentaire des TDR
- Proposition de la Méthodologie
-Visites préliminaires
-Elaboration du guide d'entretien
PHASE 3
-Travaux de terrain (enquêtes socio-économiques, relevées, mesures)
-Diagnostic des aménagements et du fonctionnement des équipements
-Proposition des solutions
PHASE 4
-Synthèse des informations
-Analyse et traitement des données
-Recommandations
-Rédaction et validation solutions
-Rapport du mémoire
FIN

CHAPTER 6

VI. GENERAL INFORMATION ON MOTOR PUMPS

VI. 1. Definition and presentation of the material

A motor-driven irrigation pump is an agricultural machine consisting of a heat engine connected to a centrifugal pump that is used to draw water from a given point (a source) and deliver it to a plot to be irrigated. The most common type of motor pump used in agriculture by fruit and vegetable growers is the centrifugal pump. They come in all sizes, from small to large. Depending on the pump's drive mode, there are two (2) types:

1°) Pumps driven by pulleys and belts; the link between the pump and the motor is ensured by a layer of V-belts connecting the motor pulley to that of the pump;

2°) Direct drive motor pumps, the connection between the pump and the motor is ensured by a shaft.

The small pumps have a compact, one-piece design. Medium power pumps have a flexible coupling: the two (2) shafts are placed opposite each other in a plastic sleeve which absorbs shocks due to alignment errors. On high-powered pumps, a direct coupling is used, also elastic, of the "crabot" type. This type of connection allows a smooth start-up and authorizes a misalignment margin.

VI. 2. Technical and technological description of the pumps

As its name indicates, the motor pump is a piece of equipment that has two (2) elements: a motor and a pump; each with its own characteristics.

VI.   2.1. The Engine

It has some characteristics that you should know for a better use and a good maintenance. The types of motors generally mounted on motor pumps are :

- the gasoline engine; very often mounted on low power motor pumps (Power equal to about 5.5 HP and flow rate up to 60 m^3 /h);

- The diesel engine; the most common; its power varies from 7 to 8 HP or more. It can have up to 6 cylinders and a flow rate of over 60 m^3 /h.

Photo 1: motopompe avec moteur à essence   Photo 2: motopompe avec moteur diesel

VI.2.2. *The pump*

The centrifugal pump is a hydraulic machine that uses the centrifugal force provided by the impeller to set the water in motion. It is composed of the following elements:

- the impeller, also called turbine or rotor. It is semi-open and receives its rotational energy from the shaft and transmits part of this energy to the liquid (water). This transmission of energy is done through the blades or vanes;

- The pump body; which receives the liquid coming out of the impeller and directs it to the pump's discharge port. The upper part of the body contains the priming and degassing port, and the lower part contains a draining port;

- The suction casing; integral with the pump casing. Its role is to direct the water coming from the suction pipe towards the impeller inlet;

- Accessories; which are devices put in place by the manufacturer for leakage, axial thrust on the impeller and sealing at the shaft passage: seals, gland, bearings...

VI. 3. **Typology and technical characteristics of the motor pumps surveyed**

From the observation of motor pump models used by farmers and from the survey of motor pump suppliers and traders in the city of Ouagadougou, we found a wide range of pumping equipment.

Thus, the following tables give us the different brands of motor pumps listed with their technical specifications:

Table 2: Brands and characteristics of the motor pumps surveyed.

Brands	Models	Energy	Powers		Speed (rpm)	Flow rate (m^3/h)	HMT (m)	0 Suction/ Backflow
			CV	kW				
GDI (DTE[i])	BT 100-18	diesel	12,0	8,83	2000,0	100,0	18,0	5" x 4"
HONDA	WP 30 X	gasoline	5,5	4,0	3600,0	60,0	30,0	4" x 4"
JINLING	JL 30 PG	gasoline	4,0	3,0	3600,0	56,0	30,0	76/76 mm
JUMBO		gasoline	6,0	4,4	3600,0	36,0	30,0	80/60 mm
KAMA	KDP 40	gasoline	8.5	6,3	3600,0	40,0	16,0	100/110mm
PETER-LISTER	TV 1	diesel	8,0	5,88	1800,0	94,0	20,0	100/110mm
RAJ MOT		diesel	6,5	4,78	1500,0	80,0	16,0	4"x 4"
RHINO		diesel	8,0	5,88	1500,0	100,0	18,0	5" x 5"
ROBIN	SE 80 X	gasoline	5,0	3,7	4000,0	56,0	26,0	76/76 mm
YAMAHA	YP 30 GN	gasoline	5,0	3,7	3600,0	54,0	26,0	4" x 4"
YANMAR	TF70 HSK	diesel	6,0	4,4	1500,0	85,0	22,0	100/100 mm

Table 3: Technical specifications of the motor pump models proposed by DTE

Brand	GDI	GDI	GDI	GDI	GDI
Model	DTE 60 - 30	DTE 100- 16	DTE 125 - 18	DTE 150 - 18	DTE 200 - 20
Irrigable surface	3 to 5 ha	7 to 9 ha	9 to 11 ha	11 to 13 ha	15 to 17 ha
MOTORS					
Power	6,5 HP	12 CVS	15 CVS	18 CVS	22 CV
Number of cylinders	1,0	1,0	1,0	1,0	1,0
Energy	gasoline	diesel	diesel	diesel	diesel
Speed	3600 rpm	2200rpm	2200 rpm	2200 rpm	2200 rpm
Reservoir	3.7 liters	12 liters	15 liters	15 liters	20 liters
Air filtration	dry air	oil bath	oil bath	oil bath	oil bath
start	launcher	crank handle	crank handle	crank handle	crank handle
Cooling	air	water	water	water	water
PUMPS					
Type	centrifugal	centrifugal	centrifugal	centrifugal	centrifugal
Flow rate	60 m^3/h	100 m^3/h	125 m^3/h	150 m^3/h	200 m^3/h
0 Suction	80 mm	125 mm	125 mm	150 mm	150 mm
0 Discharge	80 mm	100 mm	100 mm	125 mm	125 mm

HMT	30 m	16 m	18 m	18 m	20 m
Coupling	direct	elastic	elastic	elastic	elastic

NB: Each DTE motor pump is delivered with a set of accessories including

- 6 m of iron reinforced rubber suction hose;

- Strainer, valve and fittings ;

- At least 10 m of spiral rubber discharge hose;

- Operation and maintenance manual containing the operating curves (flow height curve, power consumption curve, efficiency curve and NPSH requirement curve);

- Spare parts catalog ;

- 1 spare fuel filter (diesel) ;

- 1 spare oil filter;

- 1 spare oil filter;

- 2 clamps, one for the suction and the other for the discharge;

- 2 spare clamps;

- 1 set of keys for the maintenance of the motor pump ;

Photo 3: One of the motor pumps delivered by the DTE

CHAPTER 7

VII. **THE MOTOR PUMP TRADE**

VII. 1. <u>The market and marketing channels</u>

There are currently a large number of motor pumps brands on the Burkinabe market from several parts of the world (Japan, China, France, England, Thailand, Dubai, India, Ghana, Nigeria, etc.).

These materials are put on the market by two types of traders:

1) Traders in the informal sector, who distribute small motor pumps of all brands purchased on international markets and which they resell at low prices. They do not provide any maintenance service or even spare parts supply.

2°) traders in the formal sector, who are increasingly specializing in this type of equipment. They attract customers by providing after-sales service.

In Burkina Faso, the main distributors of these motor pump sets are increasingly DIACFA, COBODIM, DTE, CFAO, BURKINA EQUIPEMENTS, etc.

They distribute different brands of motor pumps (PETER LISTER, YAMAHA, HONDA, LOMBARDINI, GDI, RHINO, etc.), provide installation, training of mechanics and supply of spare parts.

The supply channels are twofold: direct and traditional import channels on the one hand, and channels from English-speaking African countries on the other (Ghana and Nigeria are the hubs). Burkina Faso borders Ghana, which produces and distributes pump sets such as UPKAR and LION and spare parts of all brands that enter Burkina through informal channels at very low prices. They are available from small traders, at the market and in informal distribution channels.

Outside of these channels, motor pumps have also spread through aid projects and all forms of government support: this is the case, for example, with the Program for the Promotion of Small Village Irrigation (PPIV), which, as part of its policy, supports the rural world by equipping producers with drainage equipment, in particular motor pumps that they obtain on credit.

CECOMA, a design office in mechanical engineering and also a technical partner of the rural world, also participates in the development of irrigation through its various services, namely

-    construction and rehabilitation of motor pumps;

- the supply of spare parts;

- maintenance.

The table below shows the main distributors of pump sets in the formal sector in Burkina.

Table 4: Approved distributors of motor pumps in Burkina Faso.

Brands of motor pumps	Authorized distributors
GDI	DTE HOUSE (based in BOBO)
GRUNDFOS	BURKINA EQUIPMENTS
HONDA	DIACFA ACCESSORIES
KAMA	DIACFA ACCESSORIES
LOMBARDINI	DIACFA MATERIAUX
PETER LISTER	COBODIM
RHINO	2 Approved Merchants
YAMAHA	CAD/CAM

VII.2 How farmers acquire motor pumps

Our survey reveals that the farmer owns his motor pump, which he has always bought new. Private farmers purchase their motor pumps on their own and pay for them in cash through formal and informal channels. In the case of producer groups, credit is sometimes granted either by supply and marketing cooperatives or by projects with repayment periods of up to three years.

The fact that most purchases are made on a cash basis and outside of the projects shows that the distribution of motor pumps is not due to a deliberate and organized promotion by the development agencies. While it is true that in some sites the presence of motor pumps is due to a project credit or donation, this situation remains marginal on the whole.

VII.3 Purchase costs of motor pumps

According to information gathered from traders and distributors, the prices of small motor pumps of all brands vary between CFAF 175,000 and 450,000 in the informal sector, and between CFAF 380,000 and 750,000 in the formal sector. In the category of large motor pumps, these prices go up to 2,500,000 FCFA.

These costs, in the informal sector, do not include those of accessory materials such as pipes, valves, strainers, elbows, fittings, etc., which require significant expenses that can reach 200,000 FCFA. For some operators, these prices can equal that of the motor pump.

The prices of the main suppliers of pump sets in Burkina Faso vary according to the flow rates and power of the motors.

From the results of the surveys, we have established the following tables:

Table 5: Purchase costs of motor pumps in the formal sector.

Brand	Model	Flow rate (m^3/h)	Power	HMT (m)	Purchase price including VAT
DTE	60 - 30	60,0	6,5 HP	30,0	500.000 FCFA
DTE	100 - 16	100,0	12 CVS	16,0	1.330.000 FCA
DTE	125 - 18	125,0	15 CVS	18,0	1.450.000FCFA
DTE	150 -18	150,0	18 CVS	18,0	1.950.000FCFA
DTE	200 - 20	200,0	22 CV	20,0	2.500.000FCFA
HONDA	WP80 X	60,0	6,5 HP	30,0	500.000 FCFA
KAMA	KDP 40	40,0	6,3 KW	16,0	550.000FCFA
RHINO		80,0	8 CVS	18,0	750.000 FCA
UPKAR		80,0	8 CVS	16,0	1.285.295 FCFA
YAMAHA	YP 30 GN	54,0	5 CVS	26,0	380.000FCFA

Table 6: Purchase costs of motor pumps in the informal sector.

Brand	Model	Flow rate (m^3/h)	Power	HMT (m)	Purchase price at Account
HONDA	WP30 X	60,0	2,8 KW	26,0	175,000 FCFA
HONDA	WB 30 XT	66,0	5,5 HP	26,0	210.000FCFA
HONDA	WP20X	36,0	3.8 HP	26,0	175,000 FCFA
ROBIN	SE 80 X	56,0	5 CVS	26,0	200.000 FCA
KAMA	KDP 40	40,0	6,3 KW	16,0	450.000 FCFA
YAMAHA	MT 110	100,0	12 CVS	16,0	1.330.000 FCA

VII.4. <u>Comparison between marketing systems and models</u>

The differences between the two marketing systems are related to the quality of the equipment, the models of the different brands and the reality of the after-sales service offered to customers.

In the formal trade, there are dealers of well-known brands from the manufacturers (HONDA, YAHAMA, RHINO, GDI, GRUNDFOS...). They provide several services such as

-   the sale with guarantee of quality materials;
-   supply of the necessary accessories and original spare parts;

- Installation of pump units and training of pump operators in maintenance and upkeep;

On the other hand, in the informal sector, there are small traders who deliver pumping equipment of all brands and from everywhere! These materials are sold without any guarantee and without accessories or spare parts. The prices of motor pumps are not the same in the two marketing systems for the same brand: the differences are due to the VAT (18%) and to the fixed charges that the concessionary companies bear.

For farmers, there are no models that are better endorsed than others. The problems raised have always been general. The few differences that we have seen seem to be related to engine power, not to models and brands. We have met farmers who have experimented with several brands in their activities and they are all satisfied with their motor pumps, since for them it is more a question of achieving the desired objective, i.e. bringing the dose to the crops from the germination phase to harvest.

CHAPTER 8

VIII. DIAGNOSTIC ANALYSIS AND PERFORMANCE

VIII.   1. <u>State of operation of motor pumps</u>

VIII. 1.1. *Description of the facilities*

On the sites visited, the motor pumps are installed as close as possible to the source and exposed either completely to the sun or protected by shade and removed at the end of the activities. The exception is the sites around Lake Bam in Kongoussi, where the motor pumps are also installed in buildings built upstream of the perimeter in the lake bed. The buildings housing the motor pumps have only one door as an opening. Ventilation is inadequate and poses a problem for air-cooled motors. In order to allow for proper installation of the strainer, intake channels are dug to continuously conduct water from the water body into the wells and thus compensate for the withdrawal of water as the dry season progresses, as the near edge of the water is very muddy and this would require the permanent rearrangement of a water hole for the strainer (its installation in a vertical position requires a minimum water height of 40 cm).

Photo 4: Station de pompage sur le lac BAM

Photo 5: Motopompes exposées au soleil

VIII. *1.1.1. The pipes*

- Vacuuming

The suction lines are fixed and short (approx. 6 to 8 m). They are formed by a 90° elbow to a foot valve with a profiled vein by a rigid pipe. The diameters encountered vary from 80 to 125 mm.

- Backflow

In most cases, the discharge lines do not have a check valve. These conditions expose the pump to the phenomenon of coupling in case of sudden motor stop. The PVC installations are exposed to solar radiation and premature aging. On the Kongoussi sites, they are buried.

Connections are often poorly made, resulting in considerable leakage in the networks.

Photo 6: Canal d'amenée d'eau

Photo 7: Raccordement des conduites

VIII. *1.2 Status of the motor pumps*

From the field observations, it was found that most of the motor pumps are newly acquired, although some farmers still have old motor pumps. Some of the farmers take care of their equipment because they are careful not to use it during the hottest part of the day, especially in the hot season. Others do not take proper care of their equipment, and few build shelters, which would be useful.

The average age of the motor pumps surveyed is about four (4) years. The majority of operators are first-time users and claim to be starting out on their own, using manual dewatering with buckets and then hand or treadle pumps.

The acquisition of a new motor pump represents a further stage in the development of their farms due to the increase in marketing capacity that has led to a large number of applicants for mechanized irrigation.

VIII.1.3. Operation of the motor pumps on the sites

The pumps are designed to pump surface water and groundwater into wells. Pumping from surface water is the type of pumping encountered at all sites. Motor pumps are also installed in wells, but all wells operated have problems with drying up in the dry season.

The major problem with these wells is the low flow rate. The motor pumps empty the wells quickly and the farmer is forced to limit the flow of his motor pump or to pump intermittently.

Finally, it should be noted that on the sites studied, the motor pumps are always idling and the reasons given by the farmers are :

- extend the life of the engine;

- do not overheat the engine;

- to limit the flows so as not to destroy the canals and to have a quantity of water easy to control;

- reduce fuel consumption;

- limit the flow when you have old pipes.

VIII. 1.4. *Problems encountered*

VIII.1.4.1. *Breakdowns: nature and frequency*

Breakdowns are infrequent in most cases, as only a small number of farmers surveyed report having had breakdowns during this season. The most common breakdowns cited by farmers were for :

- change of the piston ;

- ring wear ;

- the injection pump ;

- Shirt;

- the pump seal (stuffing box);

- defusing;

- and descaling, which in our opinion is a maintenance operation.

These breakdowns pose problems for farmers because the machine can be out of action for one or two days, or even a week, for those who cannot find parts locally, as they often have to send for them in Ouagadougou or in neighboring countries (Togo or Ghana). Otherwise, farmers manage to rent a motor pump, borrow from a neighbor or relative, or at worst limit the area cultivated.

From our observation, there is no direct relationship between the number of breakdowns and the age of the engine. Some old engines run without breakdowns while others in their first year of operation have suffered a breakdown during the campaign. The frequency of intervention by mechanics is difficult to estimate. It depends on several factors, notably the age of the equipment and above all the seriousness with which it is maintained. This leads us to believe that the control of maintenance is one of the essential factors determining the frequency of breakdowns.

Interviews with farmers indicate that some brands are more prone to breakdowns than others. This has not been verified at all, neither by the repairers nor by the suppliers.

VIII. 1.4.2. *Repair*

In each of the areas visited, there are mechanics that everyone calls on in case of breakdown, although they are not often specialized in pumping equipment. These mechanics provide all the basic interventions and even some experienced farmers do so. Others, on the other hand, claim to have received training in motor pump maintenance with the support of the PPIV, PERCOM and the advice of agricultural supervisors. Others tinker with parts from other engines or call on large workshops in the capital (for modifications of parts such as pistons).

VIII. 1.4.3 *Maintenance*

In most private farmers, the owner of the pump himself works in the field and maintains his machine. However, many of these farmers have not received any training in the maintenance of the motor pump and its hydraulic equipment. Most of the time, this knowledge is acquired on the job and does not lead to good results.

The maintenance of motor pumps is not properly carried out in many cases, which does not guarantee a good operation. Not all precautions are taken in the use of some motor pumps, this situation exposes them to various problems including the defuelling of the motor pump, heating of the motors. The most commonly performed maintenance operations concern:

- changing the engine ;
- cleaning the air filter ;
- the change of spark plugs which is another frequent intervention;
- Descaling of the engine.

In our survey, the frequency of emptying is more than two (2) per month, which is correct. Some do the cleaning almost every day. Each farmer uses 4 to 5 candles per season, sometimes more. Decalcifications are in all cases done at the beginning of each season.

VIII. 1.4.4. *After-sales service*

Farmers face many problems related to the availability of spare parts and mechanics capable of doing a good repair (rendering a good service) in case of breakdowns.

The observation made in the field is that there is a lot of trial and error in the choice of parts to be replaced because there are no stores selling spare parts for the various brands of motor pumps in these localities. Farmers are therefore faced with problems that often hinder their activities.

VIII. 2. Operation of the farms

VIII.2.1. *Conduct of irrigation*

VIII.2.1.1. *Method of irrigation*

On all the farms visited, irrigation is done by gravity. The farmers use the short, plugged furrow technique. At each watering, after pumping out the water, they fill the furrows. The irrigation dose is proportional to the volume of the furrow, and varies according to the height of the ridge (about 8 to 10 cm). The plants are placed on the ridges, spaced 30 to 50 cm apart. The roots capture the water that seeps into the furrows and rises into the ridges by capillary action. This easy to manage system requires, according to the farmers, the presence of a person permanently in the plots to open and close the micro-dams in order to manage the irrigation.

Photo 8: Arrosage sur la parcelle   Photo 9 : remplissage des raies

VIII.2.1.2. _Daily irrigation duration_

Irrigation times are equal to motor pump operating times because there was never any water storage. For all the zones observed, the actual irrigation times vary from 3 to 5 hours in the cold season and reach 8 to 10 hours per day in the hot season. This is also related to the area to be irrigated. The duration of the motor pump operation is much less than the time spent by the farmers. However, this time can reach 8 hours for some farmers.

In contrast, during the hot season, most farmers spend the entire day in the plots.

The graph below shows the duration of irrigation in cold and hot seasons.

Figure 3: Daily irrigation duration

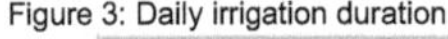

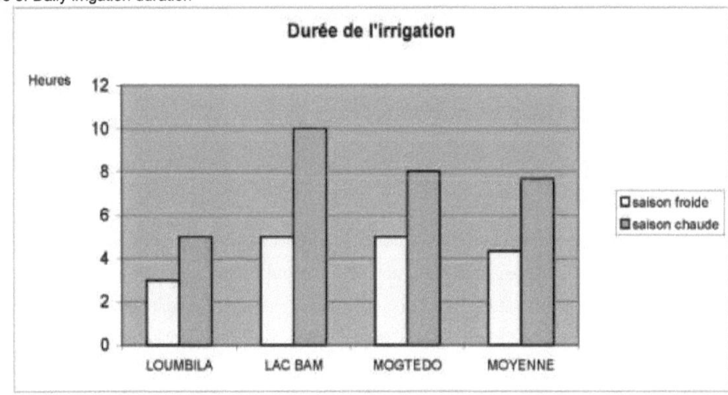

VIII. 2.1.3. _Frequency of watering_

In the hot season, farmers who grow very intensive crops on small areas irrigate almost every day. Others irrigate every 2 to 3 days or twice a week. The frequency of irrigation varies for the same farmer depending on the circumstances.

VIII.2.1.4. _Irrigation doses_

We do not have precise values for the quantities of water delivered for each irrigation. Given this lack of results, we cannot give an order of magnitude of the daily quantities supplied. Farmers irrigate with doses that they appreciate themselves. This is not in line with the theoretical recommendations. Farmers seem to have a fairly good control, except in aberrant cases, of irrigation water supplies that correspond to the needs of the crops.

VIII.2.1.5. _Soils of the perimeters_

Knowledge of the nature of a soil allows us to determine its agronomic aptitude or fertility, to understand the level of agricultural production of a cropping system and to consider improvements if possible. Two types of difficulties limited the pedological analysis of the farms visited:

- the absence of a benchmark to allow the overlay of soil maps and network plans;
- the very high cost of soil testing.

From the observation made and the information gathered from agricultural supervisors, the different soils of the farms visited are generally poor in organic matter (the case of Mogtedo is confirmed by the analyses carried out by the PMI-BF), because of the almost total exploitation or burning of straw but especially because of the low doses of organic matter.

VIII.2.1.6 _Agronomic management of perimeters_

Agriculture is an activity that uses means or factors of production: capital, labor force, agricultural equipment,...

Irrigated agriculture is perceived by farmers as an economic activity that complements traditional rain-fed agriculture. The surface areas are small, which explains the high demand. Some farmers find themselves with several scattered plots of land, making management difficult. On the whole, we note :

- the derisory state of the agricultural equipment;

-   the non respect of the cultivation calendars;

-   soil depletion;

-   the wild extension

The consequences are the disruption of the campaign and the drop in yield that has become almost recurrent on all the perimeters. In addition, the use of certain varieties of non-productive seeds.

VIII.2.1.7. *Cultivation practices on the perimeters*

On each of the farms visited, there are two (2) cropping systems, namely:

-   the irrigated crop system ;

-   the traditional rainfed farming system.

The size of farms varies greatly. Surveys conducted have shown this great variability in farm size.

In fact, family farms vary from 0.25 to 3 hectares at most, and those of producer groups up to 25 hectares.

VIII. 2.1.8. *Cultivations practiced*

The crops mainly grown on most of the perimeters are those widely consumed by the rural and urban populations, and which require an excellent marketing circuit.

These are: corn, cabbage, eggplant, green bean, onion, zucchini, okra, potato, coumbas (local eggplant), sorrel...

Photo 10: Culture de courgette   Photo 11: Culture de maïs

In addition to these crops, there is considerable diversity: other crops are grown, but only on small areas. In one area in the village of Kombango in Kongoussi, there is a combination of banana, onion and tree crops (mangoes, guava, etc.).

Yields and production in market gardening are difficult to determine accurately at some sites, given that production is most often sold in small, unweighed quantities. In Mogtedo and Kongoussi, they are satisfactory for tomatoes, eggplant and okra. These results are presented in the following table:

Table 7: Yields of market garden crops in Mogtedo and Kongoussi (T/ha)

	onion	tomato	eggplant	cabbage	okra
Mogtedo	19,0	22,0	17,0	5,0	16,0
Kongoussi	8,0	17,0	24,0	-	7,5
Average	**13,5**	**18,5**	**20,5**	**5,0**	**11,75**

Source PMI-BF Report, 1995

Harvesting is done individually. Indeed, each farmer organizes himself as he wishes to carry out his operations. Some make do with available family labor, others use hired labor.

VIII.2.2. *Irrigation economics*

VIII.2.2.1. *Marketing of the products*

Market gardening is a cash-generating activity. According to our field survey, it appears that farmers who use motor pump irrigation all cultivate for sale, since the majority of their crops are marketed. This is due to the fact that they need to be able to cover operating costs (especially fuel costs). They market to large rural and peri-urban centers. These sales are made either directly or through local purchases by private traders or cooperatives.

Not all outlets are close by: in Talembika, for example, farmers travel many kilometers with their produce to reach the local market in Mogtedo. Sales are frequent, and farmers start selling as soon as possible in order to have funds to continue their activity and to benefit from the high prices at the beginning of the season. In the middle of the season, sales continue regularly in order to have continuous income and to avoid losses due to poor conservation.

In talking with farmers, it is clear that most do not know the actual amount of their income, since the proceeds from sales are continuous and immediately reincorporated into operating expenses. Those who sell to the cooperative have figures to put forward because all receipts and expenses are recorded in an account book. These receipts range from 500,000 FCFA to more than 1,000,000 FCFA per season.

VIII.2.3. *Operating costs*

VIII.2.3.1. *The cost of access to water*

Some farmers pay for the water they draw from the canal. This is the case, for example, in the Mogtédo perimeter, where these amounts are set at between 1,000 and 1,500 FCFA per hectare. But in total, this expense is not very important for the season. Those who draw from the wells (in Loumbila) must have them cleaned out, and cleaning them out costs an average of 2,000 CFAF.

These costs include fuel consumption, maintenance and upkeep. The price of gasoline in Burkina is 585 FCFA/liter, and gas oil is 575 FCFA/liter. For small motor pumps, consumption is about 0.5 liters/hour and depends on the general condition of the motor pump and the operating conditions. An hour of pumping therefore costs about 290 FCFA in fuel.

For one season, the variations in costs among farmers are very large. They depend on the flow rate of the motor pump, the irrigation method used and the fuel consumption of the engine.

Farmers change their oil a little more frequently, usually every 100 hours of operation. An oil change often consumes 1 liter of motor oil type SAE40 or SAE50 at 1000 FCFA/liter. The cost of oil changes is therefore around 9,000 FCFA per season. Those with larger engines pay up to 20,000 FCFA for oil per season.

VIII.2.3.3. <u>Maintenance and repair costs</u>

The cost of labor varies according to the locality and the nature of the breakdown. It fluctuates between 10,000 and 30,000 FCFA per intervention and according to the nature of the breakdown. In all cases, it is an agreement between the mechanic and the owner of the motor pump. Precise information on the annual expenditure on spare parts could not be obtained because it appears that the acquisition of these parts is a real problem.

VIII.2.3.4. <u>Seeds and fertilizers</u>

Some farmers buy the seeds, sometimes for large sums of up to 100,000 FCFA. Others make their own seeds and a good part of them benefit from project assistance. Farmers are all aware of the importance of fertilizer in improving soil fertility and use chemical fertilizer or local manure. Expenditures range from 20,000 FCFA to more than 100,000 FCFA per season, depending on the area. For local manure, they buy it for smaller amounts and sometimes it is free if the farmer has his own animals.

VIII.2.3.5. <u>Workforce</u>

Many plow with their own oxen or donkeys. However, some use local labor. Expenses range up to 150,000 CFAF on large farms.

Farmers rarely spend money on transporting their produce as most use carts. In terms of investments, it is sometimes necessary to count the cost of fencing: some farmers install wire fences on parts of their farms to protect them from animals. The purchase of bags for the transport of products, which amounts to 10,000 CFA francs per season, should also be noted.

VIII.3. <u>Constraints of irrigated agriculture</u>

From the analysis made in the field, it appears that four (4) essential factors limit the development of irrigated agriculture activity:

- *The first factor is the marketing of products:* this operation is difficult to carry out. The farmer's income depends on it. The very strong competition from producers who manage to saturate the markets at certain times makes this problem very acute and sometimes insoluble. Despite a continuous increase in demand, marketing remains the primary concern of farmers;

- *The second factor is the initial capital outlay required to purchase the pumps:* Many people are unable to purchase a motor pump.

- *The third factor is farm management:* both in terms of water costs and in terms of fertilizer and other input costs, profitability can be improved. When it is not optimized, the farm becomes unprofitable. Training is needed on the management aspects of the farm and on purely technical subjects: cultivated area, improvement of motor pump flows, cultivation calendar, choice of seeds, marketing....

At present, everyone is doing well, some very well, others much less well.

- *The fourth factor is water availability:* depending on the area, the drying up of wells and artificial water tables in the dry season prohibits the extension of crops on the surface or over time.

These points constitute the main blocking factors and concern the majority of farmers. There are other limiting factors, but these are specific situations that cannot be generalized:

- the availability of land, for example, is not a cause of limitation of irrigated agriculture except in Mogtedo where all the land is allocated;

- Labor is available everywhere except in Kongoussi, an important cultivation area, where it is now necessary to travel up to 30 km to hire laborers;

- Some farmers cite the problem of fencing as a serious one because they believe that without good fencing, crops are damaged by animals.

CHAPTER 9

IX. PERFORMANCE OF MOTOR PUMPS

We generally have very little information on the nominal characteristics of motor pumps, except those indicated by the manufacturers. Very often these are incomplete or even non-existent (characteristic curves). Normally, the different brands of motor pumps should be tested on the bench. However, due to lack of time and equipment, we have looked for performance data from farmers according to the way they use their pumps. However, the information collected in the field allowed us to complete the results of the tests carried out on the bench by ETSHER with the CECOMA research department.

The performances sought in the field and verified with the results on the test bench were :

- operating rates;

- fuel consumption ;

- overall performance.

They allowed us to draw the characteristic curves of some motor pumps (see appendix).

The research and evaluation of performance was an important and indispensable step in order to know if the assigned objectives were achieved. Moreover, this step allowed us to detect possible shortcomings. The analysis of the causes of the weaknesses will lead to proposals for solutions and action strategies to be implemented to improve the observed performances.

IX. 1. <u>Operating rates of the motor pumps</u>

The motor pump flow rates measured on the farms vary greatly from one farm to another and are on average between 50 and 60% of the nominal flow rates indicated on the motor pump nameplate, which do not represent the maximum flow rates of the motor pumps. These results appear to be due either to the pumps themselves (size and condition of the motor) or to pumping practices (idling). Farmers rarely use maximum speed. The flow rate of their pumps is always much lower than the nominal flow rate. One of the difficulties of the test is that the motors do not run at constant speed. The speeds change gradually over time, sometimes accelerating and sometimes slowing down (see measurement sheet in the appendix).

Figure 4: Flow Ratio

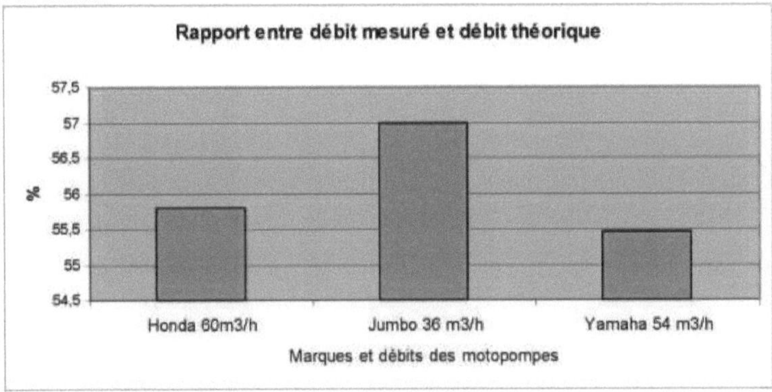

IX.　　2. <u>Fuel consumption</u>

The most common engines found among private farmers are those that run on pure gasoline, known as "dry". Diesel engines do not exist in very small sizes. As far as fuel consumption is concerned, these different types of engines do not all consume approximately the same amounts of fuel. The measured fuel consumption varies with the speed of rotation.

The different values found are shown in the appendix.

Figure 5: Fuel consumption of motor pumps (l/h)

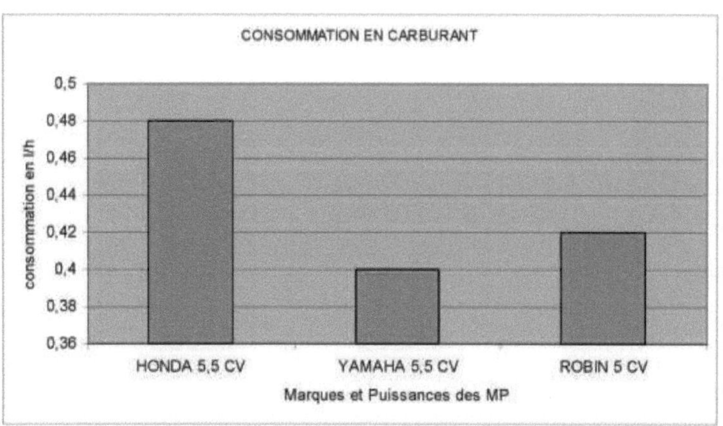

Taking into account the different densities of gasoline and gas oil, which are respectively 0.72 and 0.85 Kg/l, this should normally give us the specific consumption in grams of fuel per horsepower (g/CV). But not knowing exactly the powers provided at the time of the measurements, we cannot make these calculations.

IX.   3. **Pumping heights**

Not all motor pumps used by farmers are adapted to head (HH). We found that in most cases they are used to transport water as well as to raise it.

In fact, the pumps often work at less than 30% of their maximum head on these sites. This is too low and shows how the efficiency of the pumps is not very good. Each motor pump, depending on its design, has a set of specific characteristics depending on the speed of rotation and the head at which it operates. Unfortunately, we lack data on the characteristic curves of some of the motor pumps we have encountered. It would be interesting for us to analyze the curves and to be able to know exactly the incidence of the design of the motor pumps by the manufacturers.

Farmers' choices of motor pumps are therefore not 100% optimized. It should also be noted that many of the motor pumps sold in the informal sector have been designed abroad for needs other than market gardening. Where possible, it would be very useful to obtain the yield curves for motor pumps and to check that they correspond to the head considered.

The geometric heights studied correspond to what the surveys revealed about farmers' practices: the backflows are not very high, but because of the long lengths of the pipes, they induce high pressure losses that increase the head.

The results obtained can be found in the appendix and the figure below gives the different pumping heights according to the localities:

Figure 6: Pumping Heads

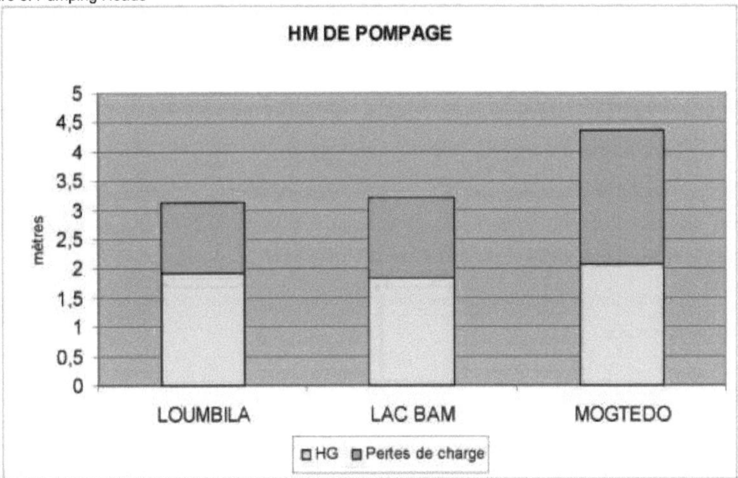

IX.   4. Yields

We wanted to have an idea of the efficiency of the motor pumps in the field by plotting the gasoline consumption as a function of the hydraulic energy returned (Q x HMT), but we can see that the efficiency is almost always the same, whatever the speed of the motor pumps.

This is surprising since it is known that the centrifugal pumps used in the engines have a decrease in performance when their speeds are reduced. There was very little data collected to draw a conclusion.

However, it would be desirable to continue such tests to determine the efficiencies of the motor pumps at different speeds.

CHAPTER 10

X.   PROPOSED SOLUTIONS

The diagnostics that we have just carried out on the operation of the pumping equipment have enabled us to gain an overall understanding of the performance of the equipment, to identify the problems and constraints, and finally to formulate proposals for improvement and recommendations for the various actors involved in the development of irrigation (farmers, cooperatives, the State, etc.).

As solutions to the problems posed, we propose:

- a sizing method that allows to choose rationally the motor pump that can bring the quantity of water according to the needs of the crop. It is a matter of precise calculations with knowledge of general and agricultural hydraulics, pumping stations;

- advice for the maintenance and good care of the equipment and also for the setting up of an after-sales service.

X. 1. <u>After sales service</u>

Farmers face serious problems related to the availability of spare parts and mechanics capable of rendering a good service in case of breakdowns. Therefore, it is very necessary to have spare parts available quickly. At this level, approved suppliers can provide this service. Thus we propose them:

to open spare parts stores in localities where irrigated agriculture is highly developed and can provide farmers with well-trained mechanics who have a perfect command of motor pump technology.

-To have technicians in Ouagadougou, ready to carry out control visits on the sites.

-To consider organizing training courses for pump operators in charge of motor pumps, this alternative will allow operators to ensure their own maintenance.

We hope that if these modest proposals come to fruition, the crucial problem of farmers will be solved. The result will be a much greater influx of farmers into the formal sector.

GENERAL CONCLUSION

The use of motor-driven pumps in irrigated areas is playing an increasingly important role in the development of irrigation. Motor-driven pumps allow significant progress in labor productivity as opposed to most drainage systems which require fixed and often expensive installations. Motor-driven pumps are very maneuverable and can operate under any circumstances.

This report brings together the results of surveys and observations on the elements of the irrigation motor pump trade. These surveys made it possible to understand the actual functioning of this equipment around the dams in Burkina Faso, to detect malfunctions, to identify the fundamental causes and finally, to formulate proposals and strategies to be implemented to improve their performance.

The overall impression from this study is that private motorized irrigation is successful, although there is still room for improvement.

We hope that our work has reached its objective, which is to make a diagnosis of the functioning of the pumping equipment and to make available to the users a method that allows them to choose their motor pump, and advice for its good functioning.

BIBLIOGRAPHY

1-   **BENHSAIN:** Pumps in rural engineering works;

2-   **BATIANA** (1991) **:** Maintenance des groupes motopompes, Mémoire de fin d'études EIER ;

3-   **GAY Bernard** (1994) : Irrigation privée et petites motopompes, CTA-GRET, 92 p.

4-   **LOMPO Tiamango** (1994): Critères de choix d'un groupe motopompe pour les périmètres irrigués, EIER end-of-studies thesis, 37 p.

5-   **CRUZ Jean François :** Les groupes motopompes sur les petits périmètres irrigués des fleuves sahéliens, Machinisme agricole tropicale N°73, 20 p.

6-   **FAO:** The pump and its maintenance, Via delle terme di carracalla ;

7-   **CHLEQ J. L, DUPRIEZ H** (1983): Les chemins de l'eau: ruissellement, irrigation, drainage, Enda, l'Harmattan, Collection Terre et Vie, 280 p.

8-   **CIRAD-CNEARC:** Méthodes d'enquêtes en milieu rural, Fascicule 3, Volume 3, 58 p.

9-   **LAMBERT Yves :** Hydraulic pumps : Possibilities of using renewable energies in Sahelian countries, CIEH

10-   **HAOUNA S.** (1995): Rentabilité financière des périmètres irrigués par pompage, Mémoire de fin d'études EIER ;

11-   **BEAUD Stéphane, WEBER Florence** (1998): Guide de l'enquête de terrain, 368 p.

12-   **DJOUKAM :** Renewable energy technology for water pumping, EIER handout

13-   **FOURNIER Jean**, (1977): Manuel de gestion des périmètres irrigués, SCET International

14-   **SALLY H, KEITA A, OUATTARA S.** (1997): Analyse diagnostic et performances de cinq périmètres irrigués autour de barrages au Burkina Faso, Rapport Final T1- PMI-BF - IIMI, 253 p.

15-   **COMPAORE M. L.** (1999): Les données de base de l'irrigation, EIER handout, 177 p.

16-   **CASTELLANET Christian** (1992): L'irrigation villageoise: Gérer les petits périmètres irrigués au Sahel. GRET, 368 p.

17-   **P. STERN ():** Small scale irrigation, A manual of low-cost water technic

18-   **CECOMA Irripro** (2003) : Guide de l'exploitation des motopompes: Instruction d'utilisation et d'entretien des motopompes diesel, 20 p.

19-   **CECOMA Irripro** (2002): Guide to the operation of motor pumps: Tips for selecting, installing, commissioning, maintaining, and repairing motor pumps, 74 pp.

20-   **DTE BF -PPIV** (2004): Guide to the use and maintenance of DTE motor pumps, 33 p.

<u>APPENDICES</u>

Appendix 1: Field Survey Form

Appendix 2: Field measurement sheet

Appendix 3: Motor pump sizing method

Appendix 4: Maintenance and servicing of motor pumps

Appendix 5: ETSHER tests on GMP

Buy your books fast and straightforward online - at one of world's fastest growing online book stores! Environmentally sound due to Print-on-Demand technologies.

Buy your books online at
www.morebooks.shop

Kaufen Sie Ihre Bücher schnell und unkompliziert online – auf einer der am schnellsten wachsenden Buchhandelsplattformen weltweit! Dank Print-On-Demand umwelt- und ressourcenschonend produziert.

Bücher schneller online kaufen
www.morebooks.shop

KS OmniScriptum Publishing
Brivibas gatve 197
LV-1039 Riga, Latvia
Telefax: +371 686 204 55

info@omniscriptum.com
www.omniscriptum.com

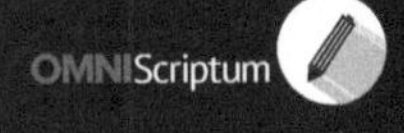

Printed by Books on Demand GmbH, Norderstedt / Germany